big & SMALL

Original Korean text by Myeong-sook Jeong
Illustrations by Mi-seon Hwang
Korean edition © Dawoolim

This English edition published by Big & Small in 2015
by arrangement with Dawoolim
English text edited by Joy Cowley
English edition © Big & Small 2015

Distributed in the United States and Canada by
Lerner Publishing Group, Inc.
241 First Avenue North
Minneapolis, MN 55401 U.S.A.
www.lernerbooks.com

ISBN: 978-1-925186-20-8

Printed in the United States of America

The Festival of the Sun

Written by Myeong-sook Jeong

Illustrated by Mi-seon Hwang

Edited by Joy Cowley

If you get lost in outer space,
look for a shining star.

6

That bright star is the sun.
Its light spreads through the solar system,
reaching our planet Earth.

When morning comes,
sunlight touches everything,
the nose of a whale,
the eyes of a child.

Living plants reach to the sky,
greeting the warm sun.

12

The sun, like a generous gardener,
gives the plants new growth.

14

As trees absorb the sunlight,
they grow taller
and their leaves get greener.

Trees take in sunlight
and give out oxygen.

Humans and animals
need oxygen to breathe.

Plants grow by absorbing sunlight
and animals grow by eating plants.
Humans obtain the energy they need
by eating plants and meat.

In spring,
the sun makes buds sprout.

In summer,
the leaves turn green.

20

In fall,
the sun ripens fruits.

In winter,
the sun melts snow.

Humans collect sunlight
to boil water and heat houses.
They also store sunlight
to light lamps in their town.

The sun makes water vapor

rise up to form clouds.

24

The clouds drift with the wind

and rain falls from them.

The sun is the source of light
and the giver of life.
While the sun is in the sky,
the whole world celebrates
the festival of growth.

28

After sunset, night comes
and we all fall asleep.
The sun will rise again tomorrow.

The Festival of the Sun

The sun is the star that provides the light and heat needed for life on Earth.

Let's think

How can plants use sunlight to grow?

What would happen if there was no sun?

What would happen if the sun got too close to Earth?

Why do the seasons change?

Let's do!

Get a thermometer and go outside and place it in a shaded area for 10 minutes.

Record the temperature.

Then place it in a sunny area for another 10 minutes.
Record the temperature again.

What is the difference in temperature?